Goldilocks and the Three Bears (Sort Of)

Adapted By John A. Honeycutt
Illustrations by Ana Nastevka
Art Direction by Kristina Ilievska

Once upon a time,
there was a little girl
named Goldilocks.

She went for a
walk in the forest.

Pretty soon,
she came upon a house.
She knocked and,
when no one answered,
she walked right in.

super-duper gravity bowl
anti-gravity bowl
regular gravity bowl

At the table in
the kitchen,
there were three
bowls of porridge.

super-duper gravity bowl
anti-gravity bowl
regular gravity bowl

A sign was
posted on each bowl.
The first sign said
"super-duper gravity bowl."
The second sign said
"anti-gravity bowl."
The third sign said
"regular gravity bowl."

super-duper gravity bowl

Goldilocks was hungry.
She tried to taste a
spoonful of porridge
from the first bowl.
"This porridge seems
too heavy,"
she exclaimed.

So, she tried to
taste the porridge
from the second bowl.

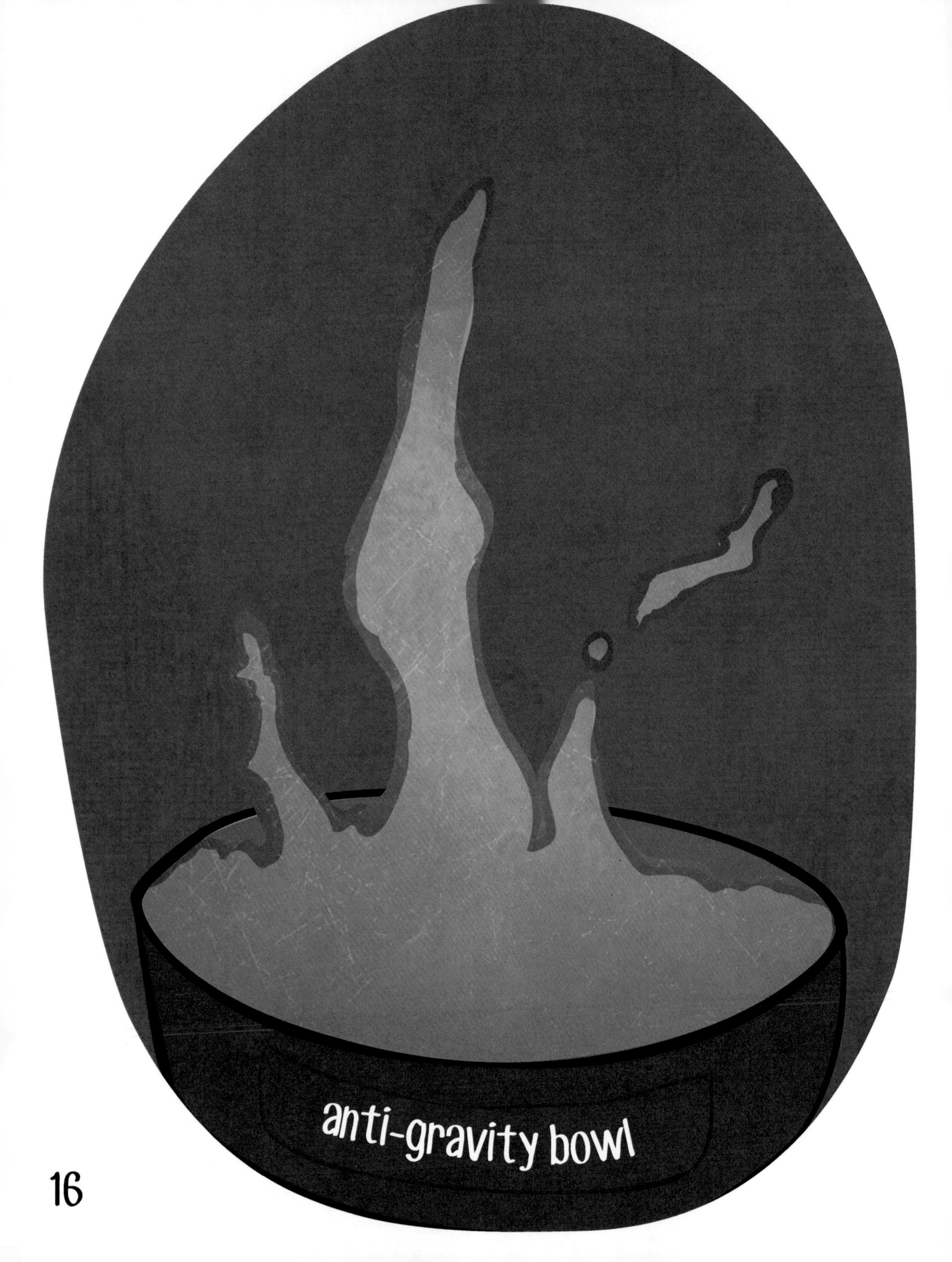
anti-gravity bowl

Before she had
a chance, the porridge
floated away.

"This porridge seems too light," she said.

REGULAR GRAVITY

So she tasted the
last bowl of porridge.
"Ahhh, this porridge
is just right,"
she said happily
and she ate it all up.

SUPER-DUPER GRAVITY BOWL
ANTI-GRAVITY BOWL
REGULAR GRAVITY BOWL

After she'd eaten
breakfast, she
decided she was
feeling a little tired.

super-duper gravity chair
anti-gravity chair
regular gravity chair

So she walked into
the living room where
she saw three chairs.

SUPER-DUPER GRAVITY CHAIR
ANTI-GRAVITY CHAIR
REGULAR GRAVITY CHAIR

A sign was posted
above each chair.

ANTI-GRAVITY CHAIR

Goldilocks sat in
the "super-duper
gravity chair"
to rest her feet.

"Oh my – this chair
pulls me down.
I can hardly get
myself out of it,"
she exclaimed.
It was very difficult
to get out of that chair.

SUPER-DUPER GRAVITY CHAIR

So she tried to sit in the second chair. Every time she tried, she floated upward – never able to actually sit down.

SUPER-DUPER GRAVITY CH
CHAIR

"This chair is
too weird!"
she whined.

ANTI-GRAVITY CHAIR
CHAIR

So she tried
the last chair.
It was the
"regular gravity chair."
"Ahhh, this chair is
just right,"
she sighed.

Goldilocks was very tired by this time, so she went upstairs to the bedroom.

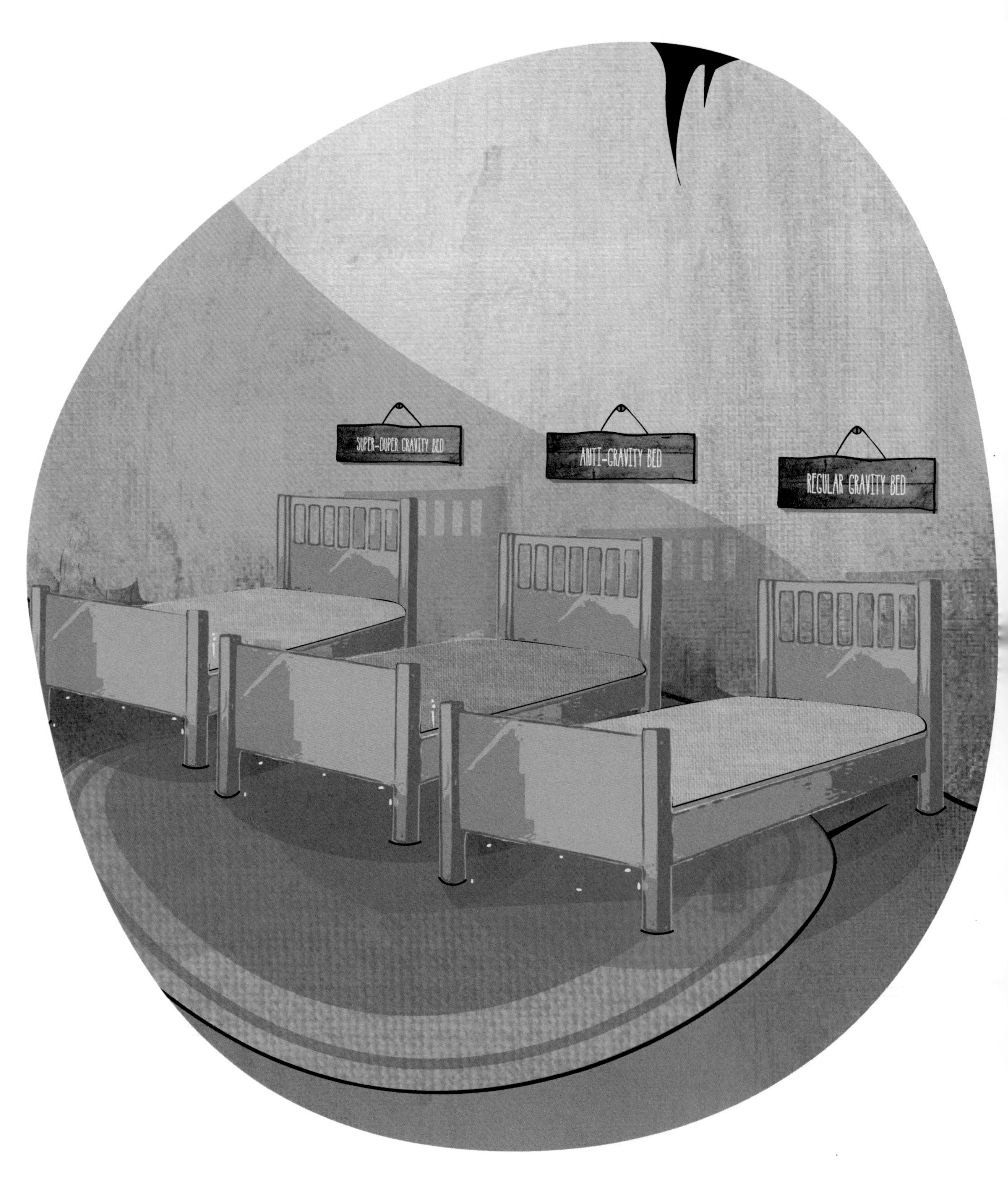
SUPER-DUPER GRAVITY BED
ANTI-GRAVITY BED
REGULAR GRAVITY BED

She found three beds.
A sign was placed
above each of them.
She didn't even try the
super-duper gravity bed.

SUPER-DUPER GRAVITY BED
ANTI-GRAVITY BED

The anti-gravity bed
was interesting, though.
When she tried to
lay down in it,
she just floated above it.

SUPER-DUPER GRAVITY BED
ANTI-GRAVITY BED
REGULAR GRAVITY BED

The third bed with
regular gravity
was just right.

As she was sleeping,
the three bears came home.
She heard them because
the noise woke her up.

SUPER-DUPER GRAVITY BOWL
ANTI-GRAVITY BOWL
REGULAR GRAVITY BOWL

The bears noticed their porridge...

SUPER-DUPER GRAVITY CHAIR
ANTI-GRAVITY CHAIR
REGULAR GRAVITY CHAIR

...and their chairs.

Before they came upstairs, Goldilocks climbed out the window and jumped to the ground.

She didn't fall too
fast or too hard.
There wasn't
super-duper gravity.
She didn't float up.
There wasn't anti-gravity.

She landed just right.
It was regular
gravity outside.

Then she ran away, never to come back to that place again.

About Gravity

Gravity is a natural, observable fact. Sometimes it is called gravitation. Gravity causes all physical things to pull toward each other. The Earth's gravity pulls objects toward it.

Gravity gives weight to physical objects. Gravity causes things to fall toward the ground when they are dropped. Gravity keeps planets in their orbit. And gravity keeps the moon orbiting around the Earth.

A very long time ago, a scientist named Isaac Newton described gravity to other scientists. His descriptions influenced a lot of other scientists to learn more about our universe. Isaac Newton was born in England in 1643. He lived to be eighty years old. Before he died, his name became Sir Isaac Newton.

There is a fun story about Isaac Newton when he was sitting under an apple tree. The story says that an apple fell on his head. When this happened, he suddenly thought of the Universal Law of Gravitation. The real story is probably more complicated than this. Either way, he was a good scientist. He was a physicist and a mathematician.

Physics is a type of science that uses a lot of math.

When studying about physics, scientists get to learn about the "four fundamental interactions in nature." One of these four interactions is gravity. Another one is a very big word—"electromagnetic force." The other two have to do with the interactions between very small particles — things that are even smaller than molecules and atoms. These two interactions are called strong nuclear force and weak force.

Gravity is easy to observe. Go throw a ball up into the air. It will drop back down to you because of gravity.

For more information, visit Hare-Brain.com.

For Parents and Teachers.

"Anti-gravity" and "Super-duper gravity" are fiction only.

To be clear, in Newton's law of universal gravitation, gravity is described as an external force transmitted by unknown means. In the 20th century, Newton's view was subsequently replaced by concepts of general relativity - whereas gravity is not viewed as a force per se - rather, gravity is the result of the geometry of spacetime.

This is important because under general relativity, anti-gravity is impossible except under fictional conditions - such as in this adaption of Goldilocks.

That said, some modern-day quantum physicists have proposed the existence of something called "gravitons" - which are theoretically a set of massless elementary particles that transmit these forces. The possibility of creating or destroying "gravitons" is unclear - but this author presently doesn't foresee the notion of "anti-gravity" becoming a serious mainstream consideration any time soon.

Hopefully, this short story inspires your young scientist to consider the concepts of gravity as a physical, observable phenomenon - and the (likely) fictional usage of "anti-gravity" and "super-duper gravity" make for a fun story without detracting from the fascinating non-fiction of present-day, established science.

Special Note about this Version of Goldilocks.

The story, concepts, names, and sequence of Goldilocks is attributed to several individuals and organizations dating to the early 19th century. Possibly, the first similar tale was written down by Eleanor Mure (1831); An oral version of the tale was provided by Robert Southey and then anonymously published (1837); George Nicol published a version of Southey's rhyme prose (1837). Joseph Cundall adapted the tale (1849); Flora Annie Steel renamed the protagonist Goldilocks (1918). Respectively, Walt Disney and Metro-Goldwyn-Mayer released animated adaptations of the tale in 1922 and 1939. This adaptation by John A. Honeycutt (2014) draws from the enchanting public domain story which has evolved over the centuries. Gratitude is extended to the tellers-of-tales, illustrators, animators, and other creative people who have each provided a dash of extraness to the porridge.